# Test Flight Saga

*The inside story on the situation of the stuck astronauts who went on a test mission to the International Space Station*

George Hoover

Copyright © George Hoover 2024

All rights reserved. No part of this book may be reproduced, stored in a retrieval system, or transmitted in any form or by any means—electronic, mechanical, photocopying, recording, or otherwise—without the prior written permission of the publisher, except for brief quotations embodied in critical reviews and certain other noncommercial uses permitted by copyright law.

# Table of Contents

Introduction

Chapter 1: The Launch of the Starliner
    A Historic Moment: The First Crewed Flight
    Meet the Astronauts: Barry Wilmore and Sunita Williams
    Chapter 2: The Journey to the International Space Station
    The Thrill of Launch: Experiencing Liftoff
    Navigating the Cosmos: The Approach to the ISS

Chapter 3: Unexpected Challenges
    Technical Glitches: Propulsion System Leaks
    Thruster Failures: Assessing the Risks

Chapter 4: Life Aboard the ISS
    Daily Routines: Work and Research in Space
    The Human Experience: Coping with Isolation

Chapter 5: NASA's Response to the Crisis
    Evaluating Safety: The Decision-Making Process
    Contingency Planning: Exploring Alternative Options

Chapter 6: The Role of SpaceX
    Rivalry in Space: Boeing vs. SpaceX
    Crew Dragon: A Backup Plan for Return

Chapter 7: The Astronauts' Perspective
    Reflections from Orbit: Wilmore and Williams

- Speak
- The Emotional Toll: Adapting to Extended Missions

Chapter 8: The Science of Space Travel
- Understanding Microgravity: Effects on the Human Body
- Research Opportunities: Experiments Conducted Aboard the ISS

Chapter 9: The Future of Space Exploration
- Lessons Learned: Implications for Future Missions
- The Next Frontier: Preparing for Mars and Beyond

Conclusion
- A New Chapter in Space History
- The Legacy of the Starliner Mission

# Introduction

The new era of space travel is upon us, characterized by unprecedented advancements, increased participation from private enterprises, and a shift in the dynamics of exploration. As we transition into what many are calling the "third space age," the landscape of space exploration is evolving rapidly, driven by technological innovation and a burgeoning commercial sector. This introduction will explore the transformative nature of this era and the foundational role of NASA's Commercial Crew Program in shaping the future of human spaceflight.

## The New Era of Space Travel

Space exploration has entered a renaissance, reminiscent of the fervor experienced during the Apollo missions. However, this time, the driving forces are not solely government agencies but also a myriad of private companies. This shift has democratized access to space, allowing a broader range of nations and organizations to participate in the exploration of the cosmos. The barriers to entry have significantly lowered, and the costs associated with launching payloads into orbit have decreased dramatically, thanks in part to advancements in reusable rocket technology.

The emergence of private space companies has fostered a competitive environment that is propelling innovation at an unprecedented pace. Companies like SpaceX, Blue Origin, and Virgin Galactic are not only launching satellites but are also paving the way for human space travel and even space tourism. The excitement surrounding these developments is palpable, as they promise to make space more accessible than ever before.

The allure of space is not just about exploration; it also encompasses the potential for economic growth. The global space economy is projected to exceed $1 trillion within the next two decades, driven by advancements in satellite technology, space tourism, and deep-space exploration. This economic potential is attracting investment from both public and private sectors, further fueling the race to explore beyond our planet.

As we delve into this new era, it is essential to recognize the historical context that has led us here. The space race of the mid-20th century, characterized by intense competition between the United States and the Soviet Union, laid the groundwork for the current landscape. The successful launch of Sputnik in 1957 by the Soviet Union marked humanity's first foray into space, igniting a series of milestones that would culminate in the Apollo moon landings. However, the end of the Cold War ushered in a more collaborative phase, epitomized by the construction of the International Space Station (ISS), where astronauts from various nations have worked together since 2000.

Despite this collaborative spirit, the past decade has seen a resurgence of competition, particularly as new players enter the field. Countries like China and India have made significant strides in their space programs, challenging the traditional dominance of the United States and Russia. This geopolitical landscape is further complicated by the involvement of private companies, which are reshaping the dynamics of space exploration.

## Setting the Stage: NASA's Commercial Crew Program

At the forefront of this new era is NASA's Commercial Crew Program, a pivotal initiative designed to facilitate the safe and reliable transport of astronauts to and from the ISS. Launched in 2010, the program aims to foster partnerships with private companies to develop crew transportation capabilities, thereby reducing reliance on foreign spacecraft for human spaceflight.

The Commercial Crew Program has made significant strides in recent years, culminating in successful crewed missions aboard SpaceX's Crew Dragon spacecraft. These missions have not only demonstrated the viability of commercial space travel but have also set a precedent for future collaborations between government agencies and private enterprises. By leveraging the expertise and innovation of the private sector, NASA is positioning itself to remain at the forefront of space exploration.

One of the key objectives of the Commercial Crew Program is to enhance safety and reliability in human spaceflight. NASA has established rigorous standards and oversight processes to ensure that commercial vehicles meet the agency's safety requirements. This collaborative approach allows for the sharing of knowledge and resources, ultimately benefiting both NASA and its commercial partners.

The successful launch of the Crew Dragon spacecraft in 2020 marked a historic milestone, as it was the first crewed launch from U.S. soil since the Space Shuttle program ended in 2011. This achievement not only restored American capabilities for human spaceflight but also signaled a new era of cooperation between NASA and private companies. The Crew Dragon's reusable design has significantly reduced launch costs, making space travel more economically viable.

In addition to SpaceX, Boeing is also a key player in the Commercial Crew Program, developing its spacecraft, the CST-100 Starliner. While the Starliner has faced technical challenges that have delayed its operational debut, NASA remains committed to supporting Boeing in resolving these issues. The partnership exemplifies the collaborative spirit of the Commercial Crew Program, as both companies work towards a shared goal of advancing human spaceflight.

The implications of the Commercial Crew Program extend beyond the immediate goals of transporting astronauts to the ISS. By fostering a competitive environment among private companies, NASA is encouraging innovation and driving down costs, ultimately making space more accessible to a wider audience. This shift has the potential to unlock new opportunities for research, exploration, and even tourism.

As we look to the future, the Commercial Crew Program is poised to play a crucial role in NASA's Artemis program, which aims to return humans to the Moon and eventually send astronauts to Mars. The lessons learned from the Commercial Crew Program will inform the development of new technologies and capabilities necessary for deep-space exploration.

In conclusion, the new era of space travel is characterized by a convergence of public and private efforts, fueled by technological advancements and economic potential. NASA's Commercial Crew Program stands as a testament to this collaborative spirit, paving the way for a future where space exploration is no longer the exclusive domain of governments but a shared endeavor among nations and private enterprises. As we embark on this exciting journey, the possibilities for discovery and innovation are limitless, promising to reshape our understanding of the universe and our place within it.

# Chapter 1: The Launch of the Starliner

## A Historic Moment: The First Crewed Flight

On June 5, 2024, a significant milestone in human spaceflight was achieved with the launch of Boeing's CST-100 Starliner spacecraft. This mission marked the first crewed flight of the Starliner, a spacecraft designed to transport astronauts to and from the International Space Station (ISS). The launch took place from Cape Canaveral Space Force Station in Florida, a site steeped in history and synonymous with space exploration.

The journey to this moment was not without its challenges. The Starliner program had faced numerous delays and technical hurdles since its inception, making the successful launch all the more poignant. Originally slated for its first crewed flight several years earlier, the mission had been postponed due to a series of technical issues, including problems with valves and helium leaks. Each setback tested the resolve of the teams at Boeing and NASA, but they remained committed to ensuring the spacecraft's safety and reliability.

As the countdown clock ticked down, anticipation filled the air. Families, friends, and colleagues of the astronauts gathered to witness this historic event. The crew consisted of two seasoned NASA astronauts: Barry "Butch" Wilmore and Sunita "Suni" Williams. Both have extensive backgrounds in aviation and spaceflight, making them ideal candidates for this pivotal mission.

The launch itself was a spectacle, a fiery ascent into the sky that captured the attention of millions. As the Starliner lifted off, it symbolized not just a technological achievement, but also a new chapter in the story of human space exploration. This mission was designed to test the capabilities of the Starliner, paving the way for future commercial spaceflight operations.

Upon successfully reaching orbit, the Starliner performed a series of maneuvers to align itself for docking with the ISS. The spacecraft's autonomous systems were put to the test as it approached the station, showcasing the advanced technology that had been developed over years of rigorous testing and evaluation. This autonomous capability is crucial for future missions, as it reduces the workload on astronauts and enhances safety during critical phases of flight.

However, the mission was not without its complications. Shortly after launch, issues arose with the spacecraft's propulsion system, including thruster malfunctions and helium leaks. These challenges prompted NASA and Boeing to closely monitor the situation, ensuring the safety of the crew while they continued their journey to the ISS. Despite these setbacks, the Starliner successfully docked with the ISS on June 6, just a day after launch, demonstrating the resilience of the spacecraft and its crew.

The arrival of Wilmore and Williams at the ISS was met with enthusiasm from their fellow astronauts already aboard the station. The ISS serves as a unique laboratory for scientific research and international collaboration, and the addition of the Starliner crew expanded the station's capabilities. Wilmore and Williams quickly integrated into the ongoing research and operations aboard the ISS, contributing their expertise to a range of scientific experiments.

As the crew settled into their new environment, they were reminded of the significance of their mission. The Starliner was not just a vehicle; it represented a crucial step toward a future where commercial spacecraft could routinely ferry astronauts to and from space. This mission was a testament to the partnership between NASA and Boeing, showcasing the potential of public-private collaborations in advancing human spaceflight.

The successful launch and docking of the Starliner were celebrated as a victory for all involved. It demonstrated the capabilities of the spacecraft and the dedication of the teams working tirelessly behind the scenes. However, the journey was far from over. The crew was initially expected to spend only a week aboard the ISS, but unforeseen technical issues would soon extend their stay, leading to a series of challenges that would test their resolve and adaptability.

## Meet the Astronauts: Barry Wilmore and Sunita Williams

Barry "Butch" Wilmore and Sunita "Suni" Williams are not just astronauts; they are seasoned veterans of space exploration, each bringing a wealth of experience and expertise to the Starliner mission. Their backgrounds in aviation and engineering, combined with their extensive training, make them uniquely qualified for the challenges of space travel.

**Barry Wilmore**, a former U.S. Navy captain, has a distinguished career as a test pilot and engineer. He earned his wings as a naval aviator and later transitioned to test flying, where he honed his skills in evaluating new aircraft and systems. Wilmore's first journey to space was aboard the Space Shuttle Endeavour in 2009, where he served as a pilot on the STS-133 mission. He later flew on the International Space Station (ISS) as part of Expedition 42, where he served as commander. His experience in both piloting and commanding space missions has equipped him with the knowledge and skills necessary to handle the complexities of the Starliner mission.

**Sunita Williams**, a retired U.S. Navy commander, is equally accomplished. With a background in mechanical engineering and a passion for aviation, Williams has made significant contributions to space exploration. She holds the record for the longest spaceflight by a woman, spending 195 days in orbit during her mission aboard the ISS. Williams is known for her extensive work on spacewalks, having conducted numerous extravehicular activities (EVAs) to maintain and upgrade the ISS. Her expertise in robotics and her ability to work effectively in high-stress environments make her an invaluable asset to the Starliner crew.

Both Wilmore and Williams have undergone rigorous training to prepare for their roles on the Starliner mission. This training included simulations of launch, docking, and re-entry procedures, as well as emergency protocols to ensure their safety in the event of unforeseen circumstances. Their ability to remain calm under pressure and work collaboratively with their fellow astronauts is a testament to their professionalism and dedication.

As they embarked on this historic mission, both astronauts expressed their excitement and commitment to the task at hand. They understood the significance of their role in advancing human spaceflight and were eager to contribute to the ongoing research and operations aboard the ISS. Their positive outlook and enthusiasm for the mission were evident in their communications with NASA and the public.

Throughout their time aboard the ISS, Wilmore and Williams engaged in a variety of scientific experiments, contributing to the broader goals of the ISS program. They worked alongside international partners, highlighting the collaborative nature of space exploration. The ISS serves as a platform for research in fields such as biology, physics, and materials science, and the contributions of the Starliner crew were integral to these efforts.

Despite the challenges they faced during their extended stay, both astronauts remained optimistic about their situation. They emphasized the importance of adaptability in space travel and expressed confidence in the capabilities of the Starliner and its team. Their dedication to their mission and their unwavering spirit served as an inspiration to those following their journey.

As the days turned into weeks aboard the ISS, Wilmore and Williams continued to embrace the unique opportunities presented by their extended mission. They participated in educational outreach activities, sharing their experiences with students and the public, and fostering interest in science, technology, engineering, and mathematics (STEM) fields. Their efforts to inspire the next generation of explorers underscored the importance of human spaceflight in shaping our understanding of the universe.

The launch of the Starliner and the subsequent journey of Wilmore and Williams represent a pivotal moment in the evolution of human space exploration. Their experiences highlight the challenges and triumphs of venturing into the unknown, and their contributions to the ISS underscore the collaborative spirit that defines modern spaceflight.

As we reflect on this historic moment, it is essential to recognize the broader implications of the Starliner mission. The successful launch and operation of the Starliner not only demonstrate the capabilities of Boeing's spacecraft but also signify a shift toward a future where commercial space travel becomes more commonplace. The collaboration between NASA and private companies like Boeing is paving the way for a new era of exploration, one that promises to expand our horizons and deepen our understanding of the cosmos.

In conclusion, the launch of the Starliner on June 5, 2024, marked a significant milestone in human spaceflight, showcasing the capabilities of the spacecraft and the dedication of its crew. Barry Wilmore and Sunita Williams, with their extensive backgrounds and unwavering commitment to exploration, embody the spirit of this new era. As they continue their mission aboard the ISS, they are not only contributing to scientific research but also inspiring future generations to reach for the stars. The journey of the Starliner is just beginning, and its impact on the future of space exploration will be felt for years to come.

# Chapter 2: The Journey to the International Space Station

## The Thrill of Launch: Experiencing Liftoff

On June 5, 2024, the world watched with bated breath as Boeing's CST-100 Starliner spacecraft prepared for its inaugural crewed flight. The atmosphere at Cape Canaveral Space Force Station was electric, filled with anticipation and excitement. This mission represented not just a technological achievement for Boeing and NASA, but also a significant milestone in the evolution of human spaceflight.

As the countdown reached its final moments, the roar of the rocket engines filled the air, signaling the beginning of a journey that would take two astronauts—Barry Wilmore and Sunita Williams—into the vastness of space. The Starliner, perched atop its Atlas V rocket, was designed to carry astronauts to the International Space Station (ISS) and back, marking a new era in commercial space travel.

The thrill of liftoff is a unique experience, one that few individuals will ever encounter. For

Wilmore and Williams, it was the culmination of years of training, preparation, and dedication. As the rocket ignited, the intense vibrations coursed through the spacecraft, a reminder of the raw power propelling them into the sky. The ascent was swift and exhilarating, with the crew experiencing several Gs of force as they broke free from Earth's gravitational pull.

As the Starliner climbed higher, the blue sky gave way to the blackness of space. The transition from the atmosphere to the void is a moment of profound significance for astronauts. It symbolizes not just a physical journey, but also a mental shift—a departure from the familiar and a step into the unknown. Wilmore and Williams were acutely aware of this transition, reflecting on the significance of their mission and the responsibilities they carried as representatives of humanity in space.

Once the spacecraft reached orbit, the crew experienced a moment of weightlessness, a sensation that was both liberating and disorienting. Floating in the cabin, they took a moment to absorb their surroundings, marveling at the advanced technology that surrounded them. The Starliner was equipped with state-of-the-art systems designed to ensure the safety and comfort of its crew, including touch-screen controls and an array of monitoring systems.

The journey to the ISS was not just about reaching a destination; it was also a test of the Starliner's capabilities. The spacecraft was designed to operate autonomously, with the ability to dock with the ISS without direct intervention from the crew. This capability is crucial for future missions, as it reduces the workload on astronauts and enhances safety during critical phases of flight.

As the Starliner approached the ISS, the crew prepared for the next phase of their mission. The spacecraft's systems were put to the test as it executed a series of maneuvers to align itself with the ISS. This phase required precision and coordination, as even minor errors could have significant consequences. The crew maintained constant communication with mission control, receiving updates and guidance as they approached their destination.

# Navigating the Cosmos: The Approach to the ISS

The approach to the International Space Station is a carefully orchestrated process, requiring meticulous planning and execution. As the Starliner closed in on the ISS, the crew relied on a combination of onboard systems and ground support to ensure a successful docking. The spacecraft's sensors and cameras provided real-time data, allowing the crew to monitor their trajectory and make necessary adjustments.

The ISS orbits Earth at an altitude of approximately 248 miles (400 kilometers), traveling at an impressive speed of about 17,500 miles per hour (28,000 kilometers per hour). This rapid pace means that the station completes an orbit around the Earth roughly every 90 minutes, presenting a unique challenge for incoming spacecraft. Timing and precision are critical, as the window for docking is limited.

As the Starliner approached the ISS, the crew focused on their instruments, monitoring the spacecraft's speed and distance. The atmosphere inside the cabin was a mix of excitement and concentration, as they prepared for the final approach. The view from the windows was breathtaking, with the Earth below appearing as a swirling blue and green orb, a stark reminder of the fragility of our planet.

The docking sequence began with the Starliner performing a series of burns to adjust its speed and trajectory. The crew communicated with mission control, who provided updates on the ISS's position and any potential obstacles. This collaboration between the crew and ground support is essential for ensuring a safe docking process.

As the spacecraft drew closer, the crew engaged the automated docking system, which took over the final approach. This system is designed to guide the Starliner into position with pinpoint accuracy, using a series of thrusters to make fine adjustments. The crew remained vigilant, ready to take manual control if necessary, but the automated systems performed flawlessly.

The moment of docking is one of the most significant in any space mission. As the Starliner made contact with the ISS, a series of latches engaged, securing the spacecraft to the station. The crew felt a sense of relief and accomplishment as they confirmed a successful docking. They had not only reached their destination but had also demonstrated the capabilities of the Starliner in a real-world scenario.

Once docked, the crew prepared to transition from the Starliner to the ISS. This process involves a series of checks and procedures to ensure that the air pressure and environment are safe for the astronauts. As they opened the hatch and stepped into the ISS, they were greeted by their fellow astronauts, who welcomed them aboard with enthusiasm.

The ISS serves as a unique laboratory for scientific research and international collaboration. It is a testament to what can be achieved when nations work together toward a common goal. The addition of Wilmore and Williams expanded the crew aboard the station, allowing for a greater range of experiments and activities.

As they settled into their new environment, Wilmore and Williams quickly integrated into the ongoing research and operations aboard the ISS. Their experience and expertise were invaluable as they contributed to various scientific experiments, ranging from studying the effects of microgravity on human biology to conducting experiments in materials science.

The journey to the ISS is not just a physical transition; it also represents a significant step in the evolution of human spaceflight. The successful launch and docking of the Starliner demonstrated the potential of commercial spacecraft to support ongoing missions to the ISS and beyond. This mission is a crucial part of NASA's broader strategy to foster partnerships with private companies, paving the way for a new era of exploration.

As the days passed aboard the ISS, Wilmore and Williams embraced the unique opportunities presented by their mission. They participated in educational outreach activities, sharing their experiences with students and the public, and fostering interest in science, technology, engineering, and mathematics (STEM) fields. Their efforts to inspire the next generation of explorers underscored the importance of human spaceflight in shaping our understanding of the universe.

The journey to the ISS is a testament to the dedication and resilience of the astronauts and the teams working behind the scenes. It showcases the incredible advancements in technology that have made human spaceflight safer and more accessible. As we look to the future, the successful launch and docking of the Starliner serve as a reminder of the potential that lies ahead.

In conclusion, the journey to the International Space Station is a complex and thrilling endeavor that requires precision, teamwork, and innovation. The successful launch of the Starliner and its approach to the ISS marked a significant milestone in the evolution of human spaceflight. Barry Wilmore and Sunita Williams exemplify the spirit of exploration, and their contributions to the ISS will have lasting implications for future missions. As we continue to push the boundaries of what is possible in space, the journey to the ISS serves as a beacon of hope and inspiration for generations to come.

# Chapter 3: Unexpected Challenges

## Technical Glitches: Propulsion System Leaks

The journey to the International Space Station (ISS) is fraught with challenges, and the inaugural crewed flight of Boeing's CST-100 Starliner was no exception. While the launch on June 5, 2024, was a historic achievement, it soon became evident that the mission would face unexpected technical challenges that would test the resilience of the spacecraft and its crew.

Shortly after launch, as the Starliner entered orbit, the crew and mission control began to receive concerning data regarding the spacecraft's propulsion system. Specifically, there were indications of leaks within the propulsion system, a critical component responsible for maneuvering the spacecraft during its journey. The propulsion system is designed to provide the necessary thrust for various maneuvers, including orbital adjustments, docking, and re-entry. Any malfunction in this system poses significant risks to the safety and success of the mission.

The initial reports of leaks were alarming. Engineers at Boeing and NASA quickly began analyzing telemetry data to determine the extent of the issue. The leaks were traced to specific valves within the propulsion system, which had not performed as expected during pre-launch tests. These valves are essential for controlling the flow of propellant, and any malfunction could compromise the spacecraft's ability to execute critical maneuvers.

As the crew continued their journey toward the ISS, they were kept informed of the situation. Barry Wilmore and Sunita Williams, both experienced astronauts, understood the gravity of the situation but remained focused on their mission. They were trained to handle emergencies and had undergone extensive simulations to prepare for a range of scenarios. Nevertheless, the reality of facing technical glitches in space added an extra layer of complexity to their mission.

The team on the ground worked tirelessly to assess the situation. Engineers conducted a thorough review of the spacecraft's systems, examining every aspect of the propulsion system to identify potential causes for the leaks. They utilized data from previous test flights and simulations to gain insights into the issue. This collaborative effort between the crew and ground support was crucial, as it allowed for real-time analysis and decision-making.

As the days progressed, the situation remained fluid. The leaks were not severe enough to pose an immediate threat to the crew, but they raised concerns about the Starliner's ability to perform necessary maneuvers, particularly during the docking process with the ISS. The crew was briefed on contingency plans, which included potential alternative strategies for docking if the propulsion system could not be relied upon.

Despite the challenges, the crew's morale remained high. Wilmore and Williams focused on their scientific objectives aboard the ISS, participating in experiments and conducting research that would contribute to the broader goals of the mission. They understood that their work was vital, not only for their mission but also for the future of human spaceflight.

The situation with the propulsion system highlighted the importance of rigorous testing and validation processes in aerospace engineering. Every component of a spacecraft must be meticulously designed, tested, and evaluated to ensure its reliability in the harsh environment of space. The Starliner program had undergone extensive testing before the crewed flight, but the complexities of space travel often reveal unforeseen challenges.

As the crew approached the ISS, the engineers on the ground continued to analyze the propulsion system's performance. They worked closely with the astronauts to monitor the situation, ensuring that all systems were functioning optimally. The collaborative spirit of the mission was evident, as everyone involved remained focused on the common goal of a successful docking.

## Thruster Failures: Assessing the Risks

In addition to the issues with the propulsion system, the Starliner faced another significant challenge: thruster failures. The spacecraft is equipped with multiple thrusters that are essential for maneuvering and stabilizing the vehicle during various phases of flight. These thrusters play a critical role in ensuring that the spacecraft can execute precise maneuvers, particularly during docking and re-entry.

As the crew prepared for the final approach to the ISS, data indicated that some of the thrusters were not responding as expected. This was a serious concern, as the ability to control the spacecraft's orientation and trajectory is vital during critical phases of flight. The crew and mission control quickly shifted their focus to assessing the risks associated with the thruster failures.

The first step was to determine the extent of the issue. Engineers analyzed telemetry data to identify which thrusters were affected and the possible causes of the failures. The team considered various factors, including temperature fluctuations, pressure anomalies, and potential software glitches. Each thruster operates independently, and understanding the specific circumstances surrounding each failure was essential for making informed decisions.

As the crew continued their approach to the ISS, they maintained constant communication with mission control. The astronauts were trained to handle such situations and were prepared to take manual control of the spacecraft if necessary. However, the reliance on automated systems is a key feature of modern spacecraft, and the crew hoped that the remaining functional thrusters would be sufficient for safe docking.

In the face of these challenges, the crew remained composed and focused. Wilmore and Williams understood the importance of teamwork and communication during critical moments. They worked closely with mission control, providing updates on the spacecraft's status and receiving guidance on how to proceed. This collaborative approach was vital in managing the risks associated with the thruster failures.

The situation underscored the complexities of space travel and the need for contingency planning. Engineers and mission planners had developed a range of backup strategies to address potential failures, and the crew was well-versed in these protocols. They reviewed their options, including the possibility of using the remaining functional thrusters to complete the docking maneuver.

As the Starliner approached the ISS, the crew executed a series of maneuvers to align the spacecraft with the docking port. The remaining operational thrusters were put to the test, and the crew relied on their training and experience to guide the spacecraft into position. The tension in the cabin was palpable as they prepared for the critical moment of docking.

The successful docking of the Starliner with the ISS was a testament to the resilience of the crew and the capabilities of the spacecraft. Despite the technical challenges, Wilmore and Williams executed the maneuver with precision, demonstrating their commitment to the mission and their ability to adapt to unexpected circumstances.

Once docked, the crew was able to breathe a sigh of relief. They had completed a critical phase of their mission, overcoming significant obstacles along the way. The collaborative efforts of the astronauts, engineers, and mission control were instrumental in achieving this success.

The challenges faced during the Starliner's journey to the ISS serve as a reminder of the inherent risks associated with space exploration. Every mission is a complex interplay of technology, human skill, and the unpredictable nature of the cosmos. The ability to respond to unexpected challenges is a hallmark of successful space missions, and the Starliner crew exemplified this spirit.

As they settled into their new environment aboard the ISS, Wilmore and Williams reflected on the journey that had brought them there. The technical glitches and thruster failures had tested their resolve, but they had emerged stronger and more determined. Their experiences would not only contribute to their growth as astronauts but would also inform future missions and advancements in space travel.

In conclusion, the Starliner's journey to the ISS was marked by unexpected challenges that tested the limits of technology and human ingenuity. The issues with the propulsion system and thruster failures highlighted the complexities of space travel and the importance of rigorous testing and contingency planning. Barry Wilmore and Sunita Williams demonstrated remarkable composure and teamwork in the face of adversity, embodying the spirit of exploration that drives humanity to reach for the stars. As they embarked on their mission aboard the ISS, they carried with them the lessons learned from their journey, paving the way for future advancements in human spaceflight.

# Chapter 4: Life Aboard the ISS

## Daily Routines: Work and Research in Space

Life aboard the International Space Station (ISS) is a unique blend of work, research, and the challenges of living in a microgravity environment. Each day for the astronauts is meticulously planned, balancing scientific duties with personal time, all while adapting to the realities of life hundreds of miles above Earth.

The astronauts typically follow a structured schedule that begins with a wake-up call around 6 a.m. Greenwich Mean Time (GMT). Despite the ISS orbiting the Earth every 90 minutes, the crew adheres to a 24-hour cycle, aligning their routines with the circadian rhythms that govern human sleep patterns. This consistency is crucial for maintaining mental and physical health in an environment where the concept of day and night is often blurred.

Once awake, the astronauts begin their morning routines, which, while similar to those on Earth, are adapted to the challenges of microgravity. Personal hygiene becomes a creative endeavor. Instead of showers, astronauts use rinseless soap and shampoo. They apply the soap, scrub, and then wipe themselves down with a towel, ensuring that no water droplets float away. Brushing teeth involves swallowing toothpaste, as spitting it out could send particles floating around the cabin. This process is a reminder of the unique adjustments required for daily life in space.

After personal care, the crew shifts focus to their work. The ISS acts as a laboratory for a wide array of scientific experiments, spanning fields such as biology, physics, and materials science. Each astronaut has specific tasks assigned to them, which may include conducting experiments, monitoring equipment, or performing maintenance on the station itself. The collaborative nature of the work fosters a strong sense of camaraderie among the crew, as they rely on one another to accomplish their goals.

A typical workday includes approximately six hours of dedicated research time. The experiments conducted aboard the ISS are invaluable for understanding how various biological and physical processes are affected by microgravity. For instance, studying the behavior of fluids in space can provide insights into fundamental physics, while research on plant growth can inform future missions to Mars and beyond. The results of these experiments contribute to a growing body of knowledge that will benefit not only space exploration but also life on Earth.

In addition to scientific research, astronauts are responsible for the maintenance of the ISS. The station is a complex machine that requires regular checks and repairs to ensure its functionality. This includes monitoring life support systems, checking for leaks, and performing routine inspections of equipment. The astronauts must be prepared to troubleshoot issues as they arise, often relying on their training and experience to resolve problems quickly.

Exercise is a critical component of daily life on the ISS. In microgravity, the human body undergoes significant changes, including muscle atrophy and bone density loss. To counteract these effects, astronauts are required to exercise for about two hours each day. The ISS is equipped with specialized exercise equipment, including a treadmill, stationary bike, and resistance machines. Astronauts strap themselves into the equipment to prevent floating away while they work out. This routine helps maintain their physical health and prepares them for the eventual return to Earth's gravity.

Meals aboard the ISS are another aspect of daily life that requires adaptation. The food is specially prepared to ensure it remains safe and nutritious in a microgravity environment. Most meals come in vacuum-sealed packages to prevent spoilage and are often dehydrated or packaged in a way that minimizes waste. Astronauts can choose from a variety of options, including fruits, nuts, and entrees that require water to rehydrate. Eating in space involves strapping meal trays to their laps or the walls to prevent food from floating away.

Meal times are also an opportunity for social interaction, as astronauts often gather to share meals and discuss their work. This camaraderie is essential for maintaining morale in the confined environment of the ISS.

As the workday comes to a close, astronauts typically have some free time before bed. This period is crucial for relaxation and mental well-being. They can engage in activities such as reading, watching movies, or communicating with family and friends back on Earth. The ability to connect with loved ones is vital for maintaining emotional health during long missions.

The day usually ends around 9:30 p.m. GMT, allowing the crew to rest and recharge for the next day's challenges. Each astronaut has a sleeping compartment, where they can secure themselves in sleeping bags to prevent floating during the night. The constant hum of machinery and air circulation systems can make sleep challenging, but many astronauts adapt to the ambient noise over time.

## The Human Experience: Coping with Isolation

Living aboard the ISS presents unique psychological challenges, primarily due to the isolation and confinement experienced by astronauts. For extended missions, the crew is cut off from the outside world, living in close quarters with a small group of individuals. This environment can lead to feelings of loneliness and stress, making coping strategies essential for mental health.

The psychological effects of isolation are well-documented in space missions. Astronauts must contend with the absence of familiar comforts and the inability to step outside for fresh air or social interactions. The confined space of the ISS means that personal space is limited, and conflicts can arise among crew members. To mitigate these issues, astronauts undergo extensive training in teamwork, conflict resolution, and stress management before their missions.

Communication with family and friends on Earth is a vital lifeline for astronauts. The ISS is equipped with internet access, allowing crew members to send emails and make occasional video calls. These interactions help maintain connections with loved ones and provide emotional support during challenging times. However, the time delay in communication can be disorienting, as messages can take several seconds to travel between Earth and the station.

To further combat isolation, astronauts are encouraged to engage in hobbies and activities that bring them joy. Personal items, such as books, photographs, and music, can be brought aboard the ISS, providing comfort and a sense of normalcy. Many astronauts take advantage of their unique vantage point to capture stunning images of Earth, sharing these moments with the public to foster a sense of connection with life on the planet below.

The beauty of Earth from space serves as a powerful reminder of home. Astronauts often report feeling a profound sense of awe and appreciation for the planet when viewing it from the ISS. This perspective can be both humbling and inspiring, reinforcing their commitment to the mission and the importance of scientific research.

Physical exercise also plays a crucial role in coping with the challenges of life aboard the ISS. The daily workout routine not only helps maintain physical health but also serves as a mental outlet for astronauts. Exercise releases endorphins, which can improve mood and reduce stress. The camaraderie built during workouts can also strengthen bonds among crew members, fostering a supportive environment.

Mental health support is a priority for space agencies, and astronauts have access to psychological resources throughout their missions. Regular check-ins with psychologists and mental health professionals help crew members address any emotional challenges they may face. These sessions provide a safe space for astronauts to discuss their feelings and experiences, ensuring that they have the support they need to thrive in the demanding environment of space.

Despite the challenges of isolation, many astronauts find their time aboard the ISS to be one of the most rewarding experiences of their lives. The opportunity to contribute to groundbreaking scientific research, work alongside international colleagues, and witness the beauty of Earth from space creates a sense of purpose and fulfillment.

In conclusion, life aboard the ISS is a complex interplay of work, research, and the human experience. Daily routines are structured to maximize productivity while addressing the unique challenges of living in a microgravity environment. Astronauts adapt their personal care, exercise, and meal preparation to suit their circumstances, all while engaging in vital scientific research that advances our understanding of the universe.

Coping with isolation is a critical aspect of life on the ISS, and astronauts employ various strategies to maintain their mental health and well-being. Communication with loved ones, engaging in hobbies, and physical exercise all contribute to a positive experience in space. As they navigate the challenges of life aboard the ISS, astronauts embody the spirit of exploration and resilience, pushing the boundaries of human knowledge and capability. Their experiences serve as a testament to the strength of the human spirit in the face of adversity, inspiring future generations to reach for the stars.

# Chapter 5: NASA's Response to the Crisis

## Evaluating Safety: The Decision-Making Process

As the Starliner mission faced unexpected challenges, NASA's response was swift and decisive. The agency's commitment to safety is unwavering, and every decision made during the crisis was driven by the need to protect the astronauts and ensure the integrity of the spacecraft. The decision-making process involved a complex interplay of data analysis, risk assessment, and collaboration among various teams and stakeholders.

At the heart of NASA's safety culture is a rigorous system of checks and balances. The agency has developed a comprehensive set of guidelines and protocols to govern every aspect of space exploration, from design and manufacturing to launch and operations. These guidelines are rooted in decades of experience and are constantly refined to incorporate lessons learned from past missions.

When the issues with the Starliner's propulsion system and thrusters emerged, NASA immediately convened a team of experts to assess the situation. This team included engineers from Boeing, NASA's Commercial Crew Program, and the agency's Safety and Mission Assurance Office. They pored over telemetry data, reviewed design specifications, and conducted simulations to identify the root causes of the problems and evaluate the potential risks.

The decision-making process was not a single event but rather a continuous dialogue among the various stakeholders. Regular briefings were held to update NASA leadership on the progress of the investigation and the potential courses of action. These briefings were attended by senior officials, including the NASA Administrator, the Associate Administrator for Human Exploration and Operations, and the Director of the Johnson Space Center.

Throughout the process, the safety of the astronauts remained the top priority. NASA's decision-making was guided by the principle of "safety first, mission second." This meant that no decision would be made that compromised the well-being of the crew, even if it meant delaying or modifying the mission objectives.

The evaluation of safety involved a multifaceted approach. Engineers analyzed the probability and potential consequences of various failure scenarios, considering factors such as the redundancy of critical systems, the ability to detect and mitigate failures, and the potential for cascading effects. They also assessed the risks associated with potential solutions, such as the impact of software updates or the use of alternative systems.

In addition to technical considerations, NASA also factored in human factors in its decision-making process. The agency recognized that the astronauts were facing significant psychological stress due to the extended mission and the uncertainty surrounding their return to Earth. Measures were taken to ensure that Wilmore and Williams received regular support from NASA's behavioral health team, including counseling and stress management techniques.

As the investigation progressed, NASA's decision-making process became increasingly complex. The agency had to weigh the risks of continuing with the Starliner mission against the potential benefits of the scientific research being conducted aboard the ISS. They also had to consider the broader implications of their decisions, such as the impact on public confidence in the Commercial Crew Program and the potential consequences for future missions.

Ultimately, NASA's decision-making process was guided by its core values of safety, integrity, and excellence. The agency recognized that the challenges faced by the Starliner mission were not unique and that similar issues could arise in future missions. As such, the lessons learned from this experience will be used to inform and improve NASA's decision-making processes going forward.

## Contingency Planning: Exploring Alternative Options

As NASA grappled with the challenges facing the Starliner mission, it became clear that the original plan for the astronauts' return to Earth might not be feasible. With the issues surrounding the spacecraft's propulsion system and thrusters, the agency had to explore alternative options to ensure the safe return of Barry Wilmore and Sunita Williams.

Contingency planning is a critical aspect of space exploration, and NASA has developed a comprehensive set of protocols to address potential emergencies and unexpected scenarios. These protocols are regularly reviewed and updated to incorporate new technologies and lessons learned from past missions.

When it became apparent that the Starliner might not be able to safely return the astronauts to Earth, NASA immediately activated its contingency planning process. A dedicated team was assembled to explore alternative options, drawing on expertise from across the agency and its commercial partners.

One of the primary options considered was the use of SpaceX's Crew Dragon spacecraft. As part of NASA's Commercial Crew Program, SpaceX had already demonstrated the capability to transport astronauts to and from the ISS. The Crew Dragon was designed to be compatible with the ISS docking ports, and its life support systems were capable of sustaining the crew for extended periods.

50

The idea of using the Crew Dragon as a contingency option was not without its challenges. The spacecraft was not originally designed to accommodate two additional astronauts, and modifications would be required to ensure their safety and comfort. Additionally, the Crew Dragon was scheduled to launch on a separate mission in September, which would need to be adjusted to accommodate the Starliner crew.

Despite these challenges, NASA's contingency planning team worked tirelessly to develop a feasible plan for using the Crew Dragon. They conducted simulations, analyzed data, and collaborated with SpaceX to identify potential solutions. The team also considered the psychological impact on Wilmore and Williams, recognizing that an extended stay aboard the ISS could take a toll on their mental health and well-being.

In parallel with the Crew Dragon contingency plan, NASA also explored other options for the astronauts' return. These included the possibility of using the Starliner in an uncrewed configuration, with the spacecraft returning to Earth autonomously. This option would require significant modifications to the spacecraft's software and systems, as well as extensive testing to ensure its reliability.

Another option considered was the use of a Russian Soyuz spacecraft. As part of the International Space Station partnership, Russia could transport astronauts to and from the ISS. However, this option was deemed less desirable due to the logistical challenges and the potential impact on the broader relationship between NASA and its international partners.

Throughout the contingency planning process, NASA remained committed to transparency and communication. Regular briefings were held to update the astronauts, their families, and the public on the progress of the investigation and the potential courses of action. The agency recognized the importance of maintaining public trust and confidence in the Commercial Crew Program and the broader space exploration efforts.

As the contingency planning process unfolded, NASA's decision-making became increasingly complex. The agency had to weigh the risks and benefits of each option, considering factors such as cost, schedule, and the potential impact on future missions. They also had to navigate the political and diplomatic implications of their decisions, particularly in the context of the Commercial Crew Program and the relationships with international partners.

Ultimately, NASA's contingency planning process was a testament to the agency's resilience and adaptability. Despite the challenges faced by the Starliner mission, NASA remained committed to finding a solution that prioritized the safety and well-being of the astronauts. The lessons learned from this experience would be used to inform and improve NASA's contingency planning processes going forward, ensuring that the agency is better prepared to respond to future crises and unexpected scenarios.

In conclusion, NASA's response to the crisis faced by the Starliner mission was a complex and multifaceted process that involved rigorous safety evaluations and the exploration of alternative options. The agency's decision-making process was guided by its core values of safety, integrity, and excellence, and it remained committed to transparency and communication throughout the process. The lessons learned from this experience will be used to inform and improve NASA's decision-making and contingency planning processes going forward, ensuring that the agency is better prepared to respond to future challenges and continue its mission of space exploration.

# Chapter 6: The Role of SpaceX

## Rivalry in Space: Boeing vs. SpaceX

The landscape of space exploration has transformed dramatically over the past decade, with significant advances driven by both government agencies and private companies. Among the key players in this new era are Boeing and SpaceX, two giants in the aerospace industry that have engaged in a competitive rivalry that has shaped the future of human spaceflight. Their contrasting approaches to technology, safety, and innovation have not only influenced their respective missions but have also had profound implications for NASA and the broader space exploration community.

Boeing, a legacy aerospace company with a long history in aviation and space, was awarded a multimillion-dollar contract by NASA in 2014 as part of the Commercial Crew Program. This initiative aimed to develop safe and reliable spacecraft to transport astronauts to and from the International Space Station (ISS). Boeing's

CST-100 Starliner was designed with a focus on safety and redundancy, leveraging decades of experience in aerospace engineering. However, the Starliner program has faced numerous setbacks, including technical glitches and delays that have raised questions about its reliability.

In contrast, SpaceX, founded by Elon Musk in 2002, has positioned itself as a disruptive force in the aerospace industry. With its innovative approach and rapid iteration of technology, SpaceX has successfully launched and returned astronauts to the ISS using its Crew Dragon spacecraft. The company's emphasis on reusability has significantly reduced launch costs and increased the frequency of missions. SpaceX's success has not only bolstered its reputation but has also put pressure on Boeing to deliver on its promises.

The rivalry between Boeing and SpaceX is not merely a competition for contracts; it represents a broader struggle for dominance in the space industry. Each company has its vision for the future of space exploration, with SpaceX aiming for ambitious goals such as Mars colonization, while Boeing focuses on ensuring safe and reliable transportation for astronauts.

This divergence in priorities has led to differing strategies and outcomes, particularly in the context of NASA's Commercial Crew Program.

As the Starliner mission progressed, the challenges faced by Boeing became increasingly apparent. Technical issues with the spacecraft's propulsion system and thrusters raised concerns about its ability to safely transport astronauts back to Earth. NASA officials, who had initially expressed confidence in the Starliner, began to reconsider their options. The prospect of relying on SpaceX's Crew Dragon as a backup plan for the return of astronauts Barry Wilmore and Sunita Williams highlighted the shifting dynamics in the rivalry.

The situation underscored the growing unease within NASA regarding Boeing's ability to deliver a reliable spacecraft. While Boeing maintained that it had sufficient data to demonstrate the safety of the Starliner, the agency's decision-making process became more complex as it weighed the risks associated with the spacecraft against the proven capabilities of SpaceX. The possibility of using Crew Dragon to bring home the Starliner crew represented a significant shift in NASA's approach and raised questions about

Boeing's future in the Commercial Crew Program.

This rivalry is further complicated by the broader implications for the space industry. As private companies like SpaceX continue to innovate and push the boundaries of what is possible, traditional aerospace giants like Boeing must adapt to remain competitive. The challenges faced by Boeing in the Starliner program serve as a cautionary tale for other companies in the industry, highlighting the importance of agility, responsiveness, and a willingness to embrace new technologies.

## Crew Dragon: A Backup Plan for Return

As NASA grappled with the challenges posed by the Starliner mission, the agency began to explore the feasibility of using SpaceX's Crew Dragon as a backup option for the return of astronauts Wilmore and Williams. This contingency plan represented a significant pivot in NASA's strategy and underscored the growing reliance on commercial partners for human spaceflight.

The Crew Dragon spacecraft, which had already demonstrated its capabilities through multiple successful missions, emerged as a viable alternative for transporting the Starliner crew back to Earth. SpaceX's track record of reliability and innovation made it an attractive option for NASA, particularly given the uncertainties surrounding the Starliner's performance.

The decision to consider Crew Dragon as a backup plan was not made lightly. NASA officials conducted a thorough assessment of the spacecraft's capabilities, examining its life support systems, docking mechanisms, and overall safety features. The Crew Dragon was designed to accommodate up to seven astronauts, allowing for flexibility in mission planning. By reallocating two seats from the upcoming Crew-9 mission, NASA could ensure that Wilmore and Williams would have a safe and reliable means of returning home.

In addition to its technical capabilities, the Crew Dragon offered a level of familiarity for the astronauts. Wilmore and Williams had already participated in training exercises with the spacecraft, which would ease the transition should they need to switch from the Starliner to the Crew Dragon. This familiarity was crucial in ensuring that the astronauts could adapt quickly to any changes in their return plans.

While the Crew Dragon presented a promising alternative, the decision to implement this backup plan was not without its challenges. The timeline for the Crew-9 mission had to be adjusted to accommodate the Starliner crew, which required careful coordination between NASA and SpaceX. The launch of Crew-9, initially scheduled for August, was postponed to allow for additional planning and preparation.

NASA's decision-making process involved extensive discussions among various stakeholders, including engineers, mission planners, and safety experts. The agency recognized the importance of transparency and communication throughout this process, ensuring that all parties were informed of the potential changes to the mission. Regular briefings were held to update the astronauts, their families, and the public on the status of the situation.

As the situation unfolded, the collaboration between NASA and SpaceX became increasingly evident. The two organizations worked closely to develop a comprehensive plan for the Crew-9 mission, ensuring that all necessary modifications were made to accommodate the Starliner crew. This partnership exemplified the spirit of collaboration that defines the Commercial Crew Program, where government and private entities work together to achieve common goals.

The potential use of Crew Dragon to bring home the Starliner crew also highlighted the importance of redundancy in space exploration. NASA's approach to human spaceflight emphasizes the need for multiple options and backup plans to ensure the safety of astronauts. This philosophy is rooted in the lessons learned from past missions, where unforeseen challenges have necessitated quick thinking and adaptability.

As the launch of Crew-9 approached, the anticipation among the astronauts and mission control grew. The prospect of returning home on a different spacecraft added an element of excitement, but it also underscored the challenges faced by the Starliner program. The situation served as a reminder of the complexities of space exploration and the need for resilience in the face of adversity.

Ultimately, the decision to consider Crew Dragon as a backup plan for the return of Wilmore and Williams represented a significant moment in the ongoing rivalry between Boeing and SpaceX. While Boeing continued to assert the safety and reliability of the Starliner, the growing reliance on SpaceX highlighted the shifting dynamics in the aerospace industry. The outcome of the Starliner mission would have lasting implications for both companies and the future of human spaceflight.

In conclusion, the role of SpaceX in the context of the Starliner mission underscores the evolving landscape of space exploration. The rivalry between Boeing and SpaceX has driven innovation and competition, shaping the future of human spaceflight. As NASA navigated the challenges posed by the Starliner program, the consideration of Crew Dragon as a backup plan for the return of astronauts exemplified the agency's commitment to safety and reliability. The collaboration between NASA and SpaceX represents a new era in space exploration, where public and private entities work together to push the boundaries of what is possible. The lessons learned from this experience will undoubtedly inform future missions and contribute to the ongoing advancement of space technology.

# Chapter 7: The Astronauts' Perspective

## Reflections from Orbit: Wilmore and Williams Speak

As the Starliner mission extended beyond its planned duration, astronauts Barry "Butch" Wilmore and Sunita Williams found themselves in a unique position. Floating aboard the International Space Station (ISS), they became not just participants in a historic mission but also observers and storytellers, sharing their experiences and insights with the world. Their reflections from orbit provided a glimpse into the realities of life in space, the challenges they faced, and the profound sense of purpose that drives astronauts in their work.

In their first news conference from the ISS, both Wilmore and Williams expressed confidence in the Starliner's ability to return them safely to Earth, despite the technical challenges they encountered. Williams articulated a sense of optimism, stating, "I have a real good feeling in my heart that the spacecraft will bring us home, no problem."

This sentiment encapsulated the spirit of resilience that characterizes astronauts; they are trained to face adversity with determination and hope.

The astronauts described their daily routines aboard the ISS, which included conducting scientific experiments, maintaining the station, and engaging in physical exercise. Each task was not just a job but a contribution to the broader goals of space exploration and scientific discovery. Wilmore and Williams emphasized the importance of their work, knowing that the data collected during their extended stay would contribute to advancements in various fields, from medicine to materials science.

Their reflections also touched on the beauty of Earth as seen from space. Wilmore shared how looking down at the planet instilled a profound sense of connection to humanity. "When you see Earth from orbit, it's hard not to feel a sense of responsibility for our planet and its inhabitants," he remarked. This perspective is often described as the "Overview Effect," a cognitive shift in awareness that many astronauts experience when viewing Earth from space. It fosters a greater appreciation for the fragility of our world and the need for collective stewardship.

In addition to their scientific duties, Wilmore and Williams took time to engage with the public, participating in educational outreach activities. They connected with students and educators on Earth, sharing their experiences and inspiring the next generation of explorers. This outreach is a vital aspect of their mission, as it helps to demystify space travel and encourages interest in science, technology, engineering, and mathematics (STEM) fields.

The astronauts' reflections also highlighted the camaraderie that develops among crew members aboard the ISS. Living and working in close quarters fosters strong bonds, and Wilmore and Williams spoke about the importance of teamwork in overcoming challenges. They recounted moments of laughter and shared experiences that helped maintain morale during their extended stay. This sense of community is essential for mental well-being, especially in the confined environment of space.

Despite their confidence and positive outlook, Wilmore and Williams were acutely aware of the emotional toll that extended missions can take. The isolation and confinement of space travel can lead to feelings of loneliness and stress, and the astronauts acknowledged the importance of coping strategies to manage these challenges. They emphasized the need for open communication with their fellow crew members and the support systems in place to address mental health.

## The Emotional Toll: Adapting to Extended Missions

Living aboard the ISS for an extended period presents unique psychological challenges that astronauts must navigate. The combination of isolation, confinement, and the demands of their work can take a toll on mental health. Wilmore and Williams, having both spent significant time in space, were well aware of these challenges and shared their strategies for coping with the emotional aspects of their mission.

One of the most significant factors contributing to the emotional toll of space travel is the separation from family and friends. For astronauts, the distance from loved ones can create feelings of loneliness and longing. Wilmore and Williams spoke candidly about their experiences, acknowledging that while they were focused on their mission, they missed the connections to their families back on Earth. Williams noted that maintaining communication with loved ones through video calls and messages was crucial for their emotional well-being. "It's important to stay connected and share our experiences with those we care about," she said.

The astronauts also highlighted the importance of establishing routines to create a sense of normalcy in their daily lives. The structured schedule aboard the ISS, which includes designated times for work, exercise, and relaxation, helps astronauts maintain a sense of purpose and balance. Wilmore described how engaging in physical exercise not only helps counteract the physical effects of microgravity but also serves as a mental outlet. "Exercise is a great way to relieve stress and stay focused," he said. The ISS is equipped with specialized exercise equipment, allowing astronauts to maintain their physical fitness and mental health.

In addition to physical activity, Wilmore and Williams emphasized the value of engaging in personal interests and hobbies. They brought books, music, and other forms of entertainment to the ISS, which provided a welcome distraction from the demands of their work. Williams mentioned that reading and listening to music helped her unwind and recharge after a long day of conducting experiments. "Finding time for ourselves is essential," she said. "It helps us stay grounded and connected to who we are."

The psychological support provided by NASA is another critical aspect of coping with the emotional toll of extended missions. Astronauts have access to mental health professionals who offer counseling and support throughout their time in space. This resource is invaluable, as it allows astronauts to discuss their feelings and experiences in a safe environment. Wilmore and Williams both expressed gratitude for the support they received from NASA's behavioral health team, emphasizing the importance of mental health in the overall success of their mission.

As they continued their work aboard the ISS, Wilmore, and Williams remained focused on their scientific objectives while also prioritizing their mental well-being. They recognized that the challenges they faced were not just technical but also deeply human. The experience of being in space, with its unique blend of wonder and isolation, required adaptability and resilience.

The emotional toll of extended missions is not limited to astronauts; it also affects their families and support networks on Earth. The families of astronauts endure their challenges, as they navigate the uncertainty of their loved ones' safety and well-being. Wilmore's wife, Deanna, shared how the church community provided support during his absence, emphasizing the importance of connection and faith during challenging times. "We lean on our community and our faith to get through this," she said.

Wilmore and Williams' reflections from orbit serve as a reminder of the profound psychological aspects of space travel. Their experiences highlight the need for comprehensive support systems that address the emotional challenges faced by astronauts and their families. As space exploration continues to evolve, understanding and

70

addressing the psychological toll of extended missions will be essential for the success and well-being of future crews.

In conclusion, the astronauts' perspective on their extended mission aboard the ISS provides valuable insights into the realities of life in space. Barry Wilmore and Sunita Williams exemplify the resilience and determination that characterize astronauts, as they navigate the challenges of isolation, confinement, and the demands of their work. Their reflections emphasize the importance of connection, routine, and support in coping with the emotional toll of extended missions. As they continue to contribute to scientific research and inspire future generations, their experiences remind us of the profound human experience that lies at the heart of space exploration.

# Chapter 8: The Science of Space Travel

## Understanding Microgravity: Effects on the Human Body

As astronauts Barry Wilmore and Sunita Williams continued their extended mission aboard the International Space Station (ISS), they found themselves at the forefront of a critical aspect of space exploration: understanding the effects of microgravity on the human body. The unique environment of the ISS, with its near-absence of gravity, provides a laboratory for studying the physiological and psychological changes that occur when the body is no longer subjected to the constant pull of Earth's gravitational field.

One of the most significant effects of microgravity on the human body is the redistribution of fluids. In Earth's gravity, fluids in the body are pulled downward, causing the lower extremities to swell and the face to appear thinner. However, in the microgravity environment of the ISS, these fluids shift upward, leading to facial puffiness and a reduction in leg volume. This fluid shift can cause a temporary increase in intracranial pressure, which may contribute to the vision problems experienced by some astronauts.

The cardiovascular system is also significantly impacted by microgravity. Without the need to work against gravity, the heart becomes less efficient, leading to a reduction in cardiac output and a decrease in blood volume. This adaptation, known as cardiovascular deconditioning, can lead to orthostatic intolerance upon return to Earth's gravity, making it difficult for astronauts to stand upright without experiencing dizziness or fainting.

Musculoskeletal deterioration is another major concern in microgravity. Without the constant stress of weight-bearing, muscles begin to atrophy, and bones lose density at an accelerated rate. Wilmore and Williams, like all astronauts, engaged in regular exercise routines to counteract these effects, using specialized equipment such as treadmills and resistance machines. However, even with these countermeasures, some degree of muscle and bone loss is inevitable.

The immune system is also affected by the microgravity environment. Studies have shown that spaceflight can lead to changes in the function and distribution of immune cells, potentially increasing the risk of infection or illness. Additionally, the stress of living in a confined space with limited social interaction can take a toll on mental health, leading to increased levels of anxiety and depression.

Despite these challenges, the microgravity environment also presents unique opportunities for scientific research. The lack of gravity allows for the study of phenomena that are difficult or impossible to observe on Earth, such as the behavior of fluids, the growth of crystals, and the development of certain types of cells. These experiments have the potential to yield insights that could lead to advancements in fields ranging from materials science to medicine.

# Research Opportunities: Experiments Conducted Aboard the ISS

The International Space Station serves as a unique laboratory for scientific research, with astronauts like Wilmore and Williams playing a crucial role in conducting experiments and collecting data. These experiments span a wide range of disciplines, from biology and physics to materials science and human physiology.

One area of particular interest is the study of the effects of microgravity on living organisms. Researchers have sent a variety of plants and animals to the ISS to observe how they adapt to the space environment. For example, studies have been conducted on the growth and development of plants in microgravity, to understand how to grow food in space to support future long-duration missions.

Animal studies have also provided valuable insights into the physiological changes that occur in microgravity. Rodents, such as mice and rats, have been sent to the ISS to study the effects of spaceflight on bone density, muscle mass, and the immune system. These studies have the potential to inform the development of countermeasures to mitigate the negative effects of microgravity on the human body.

In addition to biological research, the ISS has also been used to study the behavior of fluids in microgravity. Understanding how liquids and gases behave in the absence of gravity has implications for the design of spacecraft systems, such as fuel tanks and life support systems. Experiments have also been conducted on the growth of crystals in microgravity, which can provide insights into the structure of proteins and other molecules.

One of the most significant areas of research aboard the ISS is the study of human physiology. Astronauts like Wilmore and Williams participate in a variety of experiments designed to understand how the body adapts to the microgravity environment. These experiments include monitoring changes in bone density, muscle mass, and cardiovascular function, as well as collecting data on the effects of microgravity on the immune system and cognitive function.

The data collected from these experiments is crucial for the development of countermeasures to mitigate the negative effects of spaceflight on the human body. Exercise regimens, nutritional supplements, and even artificial gravity have been explored as potential solutions to the challenges posed by microgravity.

In addition to its scientific value, the research conducted aboard the ISS also has important implications for future space exploration. As humanity sets its sights on missions to the Moon and Mars, understanding how the human body responds to the space environment will be essential for ensuring the safety and well-being of future astronauts.

The ISS also serves as a platform for international collaboration in space research. Scientists from around the world work together to design and conduct experiments, sharing data and insights that contribute to the broader goals of space exploration. This collaborative spirit is essential for advancing our understanding of the universe and our place within it.

As Wilmore and Williams continued their work aboard the ISS, they were acutely aware of the significance of their contributions to the field of space science. Their experiences, both positive and challenging, would inform the decisions and strategies of future missions, helping to shape the future of human spaceflight.

In conclusion, the International Space Station serves as a vital laboratory for understanding the effects of microgravity on the human body and conducting cutting-edge scientific research. The experiments conducted aboard the ISS have the potential to yield insights that could lead to advancements in fields ranging from medicine to materials science. As humanity continues to explore the cosmos, the knowledge gained from the ISS will be essential for ensuring the safety and success of future missions.

# Chapter 9: The Future of Space Exploration

## Lessons Learned: Implications for Future Missions

The ongoing journey of human space exploration has been marked by significant achievements and profound lessons. As we look to the future, the experiences gained from missions like the Starliner program will shape the trajectory of space exploration and inform the strategies employed in upcoming endeavors. The challenges faced by astronauts, engineers, and mission planners provide critical insights into the complexities of space travel, emphasizing the importance of preparation, adaptability, and collaboration.

One of the most significant lessons learned is the necessity of robust contingency planning. The Starliner mission highlighted the unpredictable nature of space travel, where technical glitches can arise unexpectedly. NASA's ability to pivot and consider alternative options, such as utilizing SpaceX's Crew

Dragon for the return of astronauts, underscores the importance of having multiple pathways to ensure crew safety. This flexibility will be crucial for future missions, particularly those venturing beyond low-Earth orbit, where the stakes are higher, and the margin for error is smaller.

Furthermore, the emotional and psychological aspects of long-duration space missions have been brought to the forefront. The experiences of astronauts like Wilmore and Williams reveal the need for comprehensive mental health support systems. As missions extend in duration, the potential for isolation, stress, and fatigue increases. Future missions must prioritize the well-being of crew members, incorporating strategies for mental health management and fostering strong interpersonal relationships among team members. This focus will be essential for maintaining morale and ensuring the success of missions to the Moon, Mars, and beyond.

The importance of international collaboration has also been reinforced. The ISS has served as a platform for cooperation among nations, and this spirit of partnership will be vital for future exploration efforts. As humanity sets its sights on ambitious goals, such as establishing a sustainable presence on the Moon and sending humans to Mars, the pooling of resources, expertise, and knowledge will enhance the chances of success. Collaborative efforts can lead to shared innovations and reduce costs, ultimately benefiting all parties involved.

Moreover, the integration of commercial partners into space exploration has proven to be a game-changer. The involvement of private companies in NASA's Commercial Crew Program has accelerated the development of new technologies and increased the frequency of missions. This public-private partnership model will continue to be a cornerstone of future exploration efforts, enabling NASA to leverage the capabilities of commercial entities while focusing on its core mission objectives.

Finally, the lessons learned from past missions emphasize the importance of scientific research and technology development. Each mission provides invaluable data that contributes to our understanding of the universe and our place within it. The ongoing research conducted aboard the ISS, as well as the experiments planned for future missions, will yield insights that can inform everything from spacecraft design to life support systems. The knowledge gained will be instrumental in preparing for the challenges of deep-space exploration.

## The Next Frontier: Preparing for Mars and Beyond

As we look toward the next frontier of space exploration, the prospect of sending humans to Mars looms large on the horizon. NASA's Artemis program aims to return humans to the Moon by 2025, establishing a sustainable presence that will serve as a stepping stone for future missions to the Red Planet. This ambitious endeavor is not merely about reaching new destinations; it is about laying the groundwork for long-term human exploration of our solar system.

The Moon will serve as a testing ground for the technologies and strategies needed for Mars missions. The establishment of a lunar base camp will allow astronauts to conduct experiments, test life support systems, and refine the logistics of living and working on another celestial body. By learning how to utilize lunar resources, such as water ice, for drinking water and fuel, future missions to Mars can be better equipped to sustain human life during extended stays.

One of the critical challenges of a Mars mission is the duration of travel. Unlike the relatively short journey to the ISS, which takes only a few hours, a trip to Mars could take several months. This extended time in space presents unique challenges for crew health, including the effects of radiation exposure, muscle atrophy, and psychological stress. To mitigate these risks, future missions will need to incorporate advanced life support systems, exercise regimens, and psychological support strategies.

NASA's Space Launch System (SLS) and the Orion spacecraft will play pivotal roles in these future missions. The SLS, the most powerful rocket ever built, is designed to carry astronauts and cargo beyond low-Earth orbit. The Orion spacecraft will provide a safe and reliable means of transportation for crews traveling to the Moon and Mars. Together, these systems will enable humanity to push the boundaries of exploration and establish a sustainable presence in other worlds.

In addition to NASA's efforts, international space agencies and private companies are also preparing for the challenges of Mars exploration. The European Space Agency (ESA), Roscosmos, and other organizations are collaborating on research and technology development to support future missions. Meanwhile, private companies like SpaceX are working on ambitious plans to send humans to Mars, with Elon Musk envisioning a self-sustaining city on the planet within the next few decades.

The exploration of Mars will not only advance our understanding of the planet but also address fundamental questions about life beyond Earth. The search for signs of past or present life on Mars is a driving force behind many missions. Robotic missions, such as NASA's Perseverance rover, are already conducting experiments to search for biosignatures and collect samples for future return to Earth. These efforts will pave the way for human exploration, as astronauts will be able to build on the knowledge gained from robotic missions.

As we prepare for the next frontier, it is essential to consider the ethical implications of space exploration. The potential for contamination of other celestial bodies and the preservation of extraterrestrial environments must be taken into account. As humanity expands its reach into the solar system, the principles of planetary protection will guide our actions, ensuring that we explore responsibly and sustainably.

In conclusion, the future of space exploration is bright and filled with possibilities. The lessons learned from past missions will inform our strategies as we embark on ambitious endeavors to return to the Moon and send humans to Mars. The integration of commercial partners, the emphasis on international collaboration, and the focus on scientific research will shape the trajectory of human exploration in the coming years. As we stand on the brink of a new era in space travel, the potential for discovery and innovation is limitless, promising to expand our understanding of the universe and our place within it.

# Conclusion

## A New Chapter in Space History

The journey of Boeing's CST-100 Starliner spacecraft has been a defining moment in the history of human space exploration. This mission, marked by both triumphs and challenges, has ushered in a new era of commercial spaceflight, paving the way for a future where private companies play an increasingly vital role in the exploration of the cosmos.

The successful launch of the Starliner on June 5, 2024, was a testament to the dedication and ingenuity of the teams at Boeing and NASA. As the spacecraft soared into the sky, it carried with it the hopes and dreams of a global community eager to witness the next chapter in the story of human spaceflight. The sight of the Starliner docking with the International Space Station (ISS) just a day later was a moment of pure elation, a validation of the hard work and perseverance that had gone into making this mission a reality.

However, the journey was not without its challenges. The issues that arose with the spacecraft's propulsion system and thrusters tested the resilience of the crew and mission control. As they grappled with the unexpected, Wilmore and Williams demonstrated the qualities that define great explorers: composure, adaptability, and an unwavering commitment to their mission. Their experiences aboard the ISS, both scientific and personal, will undoubtedly shape the future of human spaceflight.

The Starliner mission also highlighted the importance of collaboration in the pursuit of space exploration. The partnership between NASA and Boeing exemplified the spirit of innovation and teamwork that has driven humanity's journey into the unknown. By leveraging the expertise and resources of both government agencies and private companies, the Starliner program has demonstrated the potential for a future where space travel is more accessible and sustainable than ever before.

As the mission draws to a close, it is clear that the Starliner has left an indelible mark on the history of space exploration. The lessons learned from this experience will inform the strategies and technologies employed in future missions, paving the way for even greater achievements. Whether it is the establishment of a permanent human presence on the Moon, the exploration of Mars, or the discovery of life beyond Earth, the Starliner mission has laid the groundwork for a future where the boundaries of human knowledge and capability are continually pushed.

## The Legacy of the Starliner Mission

The legacy of the Starliner mission extends far beyond the immediate goals of transporting astronauts to the ISS and back. This mission has catalyzed a broader transformation in the space industry, one that will shape the future of human exploration for generations to come.

One of the most significant impacts of the Starliner mission is its role in the ongoing rivalry between Boeing and SpaceX. As these two giants of the aerospace industry compete for contracts and influence, they are driving innovation and pushing the boundaries of what is possible in space travel. The challenges faced by the Starliner program have highlighted the importance of adaptability and resilience in the face of adversity, qualities that will be essential for future success.

The Starliner mission has also demonstrated the potential of public-private partnerships in advancing the goals of space exploration. By leveraging the resources and expertise of both government agencies and private companies, this mission has shown that the future of space travel lies in collaboration rather than competition. As NASA and its international partners continue to push the boundaries of human exploration, the lessons learned from the Starliner program will be invaluable in guiding their strategies and decision-making.

Perhaps most importantly, the Starliner mission has captured the imagination of people around the world. The sight of the spacecraft soaring into the sky and the stories of the astronauts aboard the ISS have inspired a new generation of explorers and dreamers. By sharing their experiences and insights,

Wilmore and Williams have shown that the journey of space exploration is not just about reaching new destinations but also about the human experience of discovery and wonder.

As we look to the future, it is clear that the legacy of the Starliner mission will continue to shape the course of human space exploration. Whether it is the establishment of a permanent human presence on the Moon, the exploration of Mars, or the discovery of life beyond Earth, the lessons learned from this mission will be essential in guiding our strategies and decision-making. And as we continue to push the boundaries of what is possible, we will always remember the courage, determination, and spirit of exploration that defined the journey of the Starliner spacecraft.

In conclusion, the Starliner mission has marked a new chapter in the history of space exploration. By demonstrating the potential of commercial spaceflight and the power of collaboration, this mission has paved the way for a future where the boundaries of human knowledge and capability are continually pushed. As we look to the stars and dream of the possibilities that lie beyond, we will always remember the legacy of the Starliner spacecraft and the brave astronauts who carried its name into the cosmos.

www.ingramcontent.com/pod-product-compliance
Lightning Source LLC
Chambersburg PA
CBHW071946210526
45479CB00002B/826